THE
YOUNG
SCIENTIST

AN INTRODUCTION TO
Observation & Discovery

THE YOUNG SCIENTIST

AN INTRODUCTION TO
Observation & Discovery

HERON BOOKS

Published by
Heron Books, Inc.
20950 SW Rock Creek Road
Sheridan, OR 97378

heronbooks.com

———————————

Special thanks to all the teachers and students who
provided feedback instrumental to this edition.

———————————

ISBN: 978-0-89-739246-4

Printed in the USA

18 July 2022

At Heron Books, we think learning should be engaging and fun. It should be hands-on and allow students to move at their own pace.

To facilitate this we have created a learning guide that will help any student progress through this book, chapter by chapter, with confidence and interest.

Get learning guides at *heronbooks.com/learningguides*.

For teacher resources, such as a final exam, email *teacherresources@heronbooks.com*.

We would love to hear from you! Email us at *feedback@heronbooks.com*.

Your YOUNG SCIENTIST JOURNAL

Scientists love to explore the world and how things in it work. They like to go new places and discover things they've never seen before.

They also like to keep track of what they find. They often fill books with notes and drawings of what they see, and include their thoughts and questions about it. These books are called *science journals*.

What's fun about a science journal is that you can use it to draw pictures or sketches of things that interest you. You can write down ideas you have about things, make maps, write down questions you have and things you want to find out more about. You might even stick in it samples of things you find—flowers, bugs, leaves, feathers, spider's webs—who knows what?

Young Scientist

JOURNAL

The learning guide that goes with this book will sometimes ask you to look at things and make notes or drawings in a journal of your own.

Whatever you put in your science journal, it will be full of your own personal discoveries. No two journals are alike.

You can use a journal like the one shown here, or you can use a notebook of your choice. You might even want to make your own science journal and use that.

Whichever type of journal you choose, it will be a place to keep drawings and notes about what you are finding out about the world and how it works.

So get ahold of a science journal, or make one, and then get going to see what you can find out. Who knows what might be waiting for you?

IN THIS BOOK

LET'S GET STARTED!

This book can be read and enjoyed just like any other. But it can also guide you toward a life-long adventure, one of discovering more and more about the world around you!

Following each chapter of this book you will find activities that encourage you to get out into the world, to look around and see what's there. Using the *Young Scientist Journal,* or a notebook of your own, you can keep track of your observations, new ideas, and discoveries.

And your adventure doesn't need to end when you finish this book. After that, you can continue having fun creating scientific adventures all your own!

Ready? Let's go!

CURIOSITY

When he was 11 years old, Richie Stachowski (stuh KOW skee) was on vacation with his family in Hawaii. He was diving underwater with his dad, excited by the unusual, colorful fish. Richie wanted to point out things to his dad, but couldn't get his attention. He wished he could talk underwater.

Richie had always been curious. He had always wanted to investigate, to know more about things and how they worked.

And he liked to imagine ways to solve problems. When he found out that nothing had been invented for talking underwater, he thought maybe he could invent something!

He was curious to know how sound works underwater. He started trying things out in his swimming pool at home. He went to a public pool and got the pool staff interested, and they let him try different things.

After some experimenting, he created a device that divers up to 15 feet apart could use to talk to one another. He called it the Water Talkie. Now people could talk back and forth underwater!

Richie used curiosity and his imagination to solve the problem. He wanted to know more. He observed, asked questions, and experimented.

And that's what scientists do—they're curious. And because they're curious they look more closely, they notice things. They ask questions. They do experiments to test their ideas. They find answers. They solve problems to make things better.

ALL SCIENCE STARTS WITH CURIOSITY, WITH WANTING TO KNOW MORE.

Thomas Edison, who invented the first light bulb, tried thousands of ideas before he got it right. When a reporter asked, "How did it feel to fail 1000 times?" Edison replied, "I didn't fail 1000 times. The light bulb was an invention of 1000 steps."

It was his interest and curiosity that kept him going. He wanted to know more.

More recently, scientist Jill Heinerth (HY nerth) was exploring underwater caves inside an iceberg in Antarctica. On one dive, she got trapped inside one of the caves. Fortunately, she was able to find her way out. On a later dive, an iceberg she'd been exploring a few hours earlier shattered and broke apart. It was a close call and many people thought she was foolish to risk her life this way.

She felt differently. She said, "People look into caves, and they see nothing but darkness, terror, fear, claustrophobia[1]. I look into a cave and I want to know what's around the next corner."

1 Claustrophobia: the fear of being in small enclosed spaces, like elevators, closets or caves.

It is Jill Heinerth's interest and curiosity that keeps her looking and learning. And her studies help us understand more and more, for example how our oceans are changing, and how changes in the weather are affecting arctic animals like polar bears.

If you are curious about something, you're off to a good start as a young scientist!

What would you like to know more about?

CURIOSITY ADVENTURE

Let's Do This!

For this activity you will need

- your curiosity
- a science journal
- a pencil

Steps

1. Look around and find an area that looks interesting to you. It might be indoors or outdoors. Explore it, and let your curiosity lead you on an adventure.

2. While you're exploring, notice some things you've never noticed before. At the same time, practice using your science journal by making notes, sketches, drawings, or other things.

3. Once you've noticed several things you never noticed before, find another area that looks interesting to you.

4. Do the same thing. Explore it, and let your curiosity lead you. Notice some things you've never noticed before and use your science journal.

5. When you're all done, tell someone about your adventures and show them what you did with your journal.

OBSERVATION

People look at things all the time. But there's a difference between just noticing something and looking at it carefully. When you pay close attention to something, when you look at it more thoroughly, you can learn something new. You can discover things you didn't already know.

For example, you might glance at the sky outside and see that it's a sunny day. But when you look more carefully, you might notice tall, dark clouds off in the distance and a funny feeling in the air. You've just discovered a big storm on the way!

When you are interested in something and want to know more about it or understand it better, you can use your observation powers to find out as much as you can about it. This may mean using many different senses.

You might look an object over from all sides. Is it large or tiny? What shape is it? You might notice its different colors. You might even cut into it, break it open, or look at it through a magnifying glass. That might be really interesting!

You might feel it with your hands to see how rough or smooth it is, or put it to your cheek to find out how warm or cool it is. Is it heavy or light? Does it feel like plastic, glass, metal or wood?

You might smell it. That will tell you something about it. Does it have a smell? Is it a strong smell, a pleasant smell, or an unpleasant one? Does it smell like anything else you know?

If it's safe, you could taste it. Is it sweet? Salty? Sour?

If possible, you might listen to it and see what you could learn from that. Have you ever tried to hear the different instruments in a band or orchestra? How about the different songs of birds, or even the sounds of their wings in flight?

There are many ways to observe something when you want to find out as much about it as you can. And that's what scientists do. They observe more closely to find out about things!

Observation is looking at or studying something carefully to find out more about it.

While diving, Jill Heinerth observed carefully and found out a lot about the changes in our oceans caused by pollution.

Richie Stachowski read about how sound works underwater. Then he observed for himself. He wanted to find out more about why divers couldn't hear one another.

Scientists use their curiosity to lead them in a direction. Then they use their observation skills to learn more.

They also learn to observe carefully for themselves, to see what they see. We can all learn from other people's observations. It's an important part of learning. But sometimes it's important to look for yourself, to see what you see.

Scientists know how to observe things carefully—and to observe them as if no one has ever done this before!

PERSISTENCE

Have you ever tried to figure out how to do something that was difficult? Did you keep working on it or did you finally just give up? You've probably experienced both!

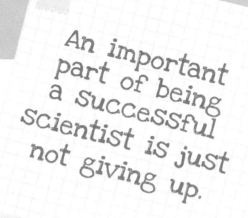

An important part of being a successful scientist is just not giving up.

Persistence is when you keep on trying, even when something is hard. Thomas Edison had persistence and because of it he was finally able to invent the light bulb.

You might be surprised to learn how important persistence is. Sometimes we might think someone is smarter or better at doing something when the person is simply more persistent. They refuse to give up.

No doctor mastered stitching up wounds on the first try.

No astronaut was an expert on moving in zero gravity without having practiced it for hours and hours first.

Two hundred years ago, a bright, curious French boy named Louis Braille (brayl) was just three years old when an accident left him blind.

Even though he couldn't see, he wanted to learn to read. At that time, the only way to do this was to trace a finger over raised alphabet letters, one by one, in special books. Louis practiced this skill for years and eventually was able to read this way.

But this skill was difficult to learn. Books written this way were very large and heavy, and there weren't many of them. Louis was curious if there could be a better way for blind people to read.

When he was twelve, he heard about a type of writing that used patterns of raised dots to stand for the letters of the alphabet. Each pattern of raised dots stood for a different letter. Originally, it had been created so soldiers could communicate with one another silently in darkness. Even though it was complicated, he decided to learn it, and after much practice, he did.

But that wasn't enough for Louis. He wanted to figure out how to help other blind people read and write. He started trying new things, changing the raised dot system, little by little, making it easier to learn and use.

Louis Braille's persistence was extraordinary. Even though he was told over and over that his ideas wouldn't work, by the time he was 15, he had created a whole new system for reading and writing by touch. Blind people no longer had to go through life without an education just because they couldn't see.

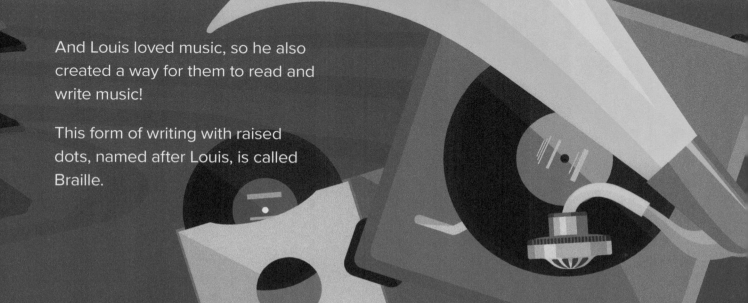

And Louis loved music, so he also created a way for them to read and write music!

This form of writing with raised dots, named after Louis, is called Braille.

Today millions of blind people are able to read and write. Because of Louis Braille's persistence, they can study, learn, and share their ideas with others through writing.

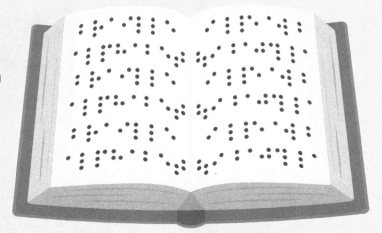

BRAILLE ALPHABET

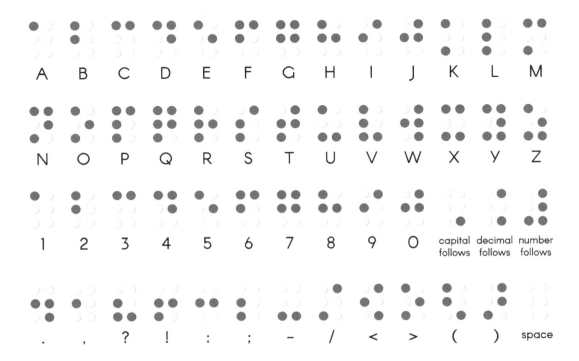

A B C D E F G H I J K L M

N O P Q R S T U V W X Y Z

1 2 3 4 5 6 7 8 9 0   capital follows   decimal follows   number follows

. , ? ! : ; - / < > ( )   space

This story teaches us one more important idea:

"LOOKING" DOESN'T JUST MEAN WITH YOUR EYES.

Louis Braille lost his sight when he was three, but he never stopped "looking." He looked with his hands and with his mind. He used the abilities he had to observe the world around him, solve problems, and help others.

OBSERVE!

For this activity you will need

- your observation powers
- your science journal
- a pencil

Steps

1. Choose one of the areas you explored before that you would like to observe more closely.

    or

    Choose a new area where you can practice your observation powers.

2. Look for things you think other people may not have noticed before. As you do this, make any notes or sketches you would like to add to your science journal.

3. When you're done with that area, choose another, either one you explored before or a new one. Look for things you think other people may not have noticed. Make any sketches or notes in your journal.

4. When you're all done, tell someone what you did during this activity, and show them what you did with your journal.

DISCOVERY

Every scientist loves making a new discovery!

But what is a discovery?

It can be as simple as finding out something new, like discovering a bird's nest you never saw before. Or it can be a more important kind of science discovery.

When a scientist makes a discovery, it usually means they've just learned something that no one ever knew before, like finding out how to make a new kind of metal, or finding a new kind of plant in the middle of a rainforest.

There's a lesson about discovering things that scientists learn over and over:

Always be willing to keep looking, to change your mind, and see what new things you can learn.

Being a scientist means being able to look at something you are familiar with in a new way. It also means being willing to change your mind if what you see is different from what you thought you would see.

In 1928, Dr. Alexander Fleming decided to take a vacation. He left behind his lab, where he was working on a cure for a disease. Upon his return, he found mold growing in a dish of diseased cells he'd been studying.

This moment turned out to be one of the most important moments in the history of human health. Dr. Fleming might have thought, "Oh, that test is wrecked because of the mold," and simply thrown it away. Instead, he remained curious about it. He figured maybe there was something there he didn't know, something new he could learn.

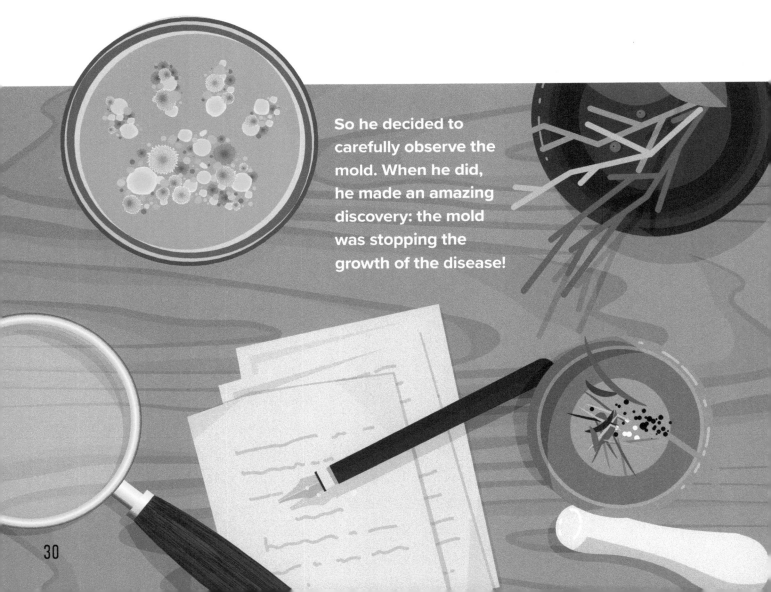

So he decided to carefully observe the mold. When he did, he made an amazing discovery: the mold was stopping the growth of the disease!

Over the next 14 years, other doctors worked to understand what had happened in Dr. Fleming's lab. Finally, a tremendously successful medicine was created from a mold. You may have heard of it: penicillin. Penicillin is a type of medicine called an antibiotic, used to cure infections.

This single discovery changed the world.
It has saved millions and millions of lives.

So if you keep observing, keep asking questions, and are always willing to look at things in new ways, you just might make some exciting discoveries of your own!

MAKE A DISCOVERY!

Let's Do This!

For this activity you will need

- your science journal
- pencil

Steps

1 Go make a discovery. Find out something new, something you didn't know about before.

2 Write down what you discovered in your science journal.

3 Share your discovery with someone.

RESEARCH

4

If you know anything about scientists, you know they do a lot of research. What does this mean exactly? What are scientists doing when they do research?

Research is the word scientists use for *studying, observing, experimenting,* and *discovering.*

When you look for information, that's research. When you ask questions and search for answers by observing things, you are researching. When you do an experiment, that's research too.

When Jill Heinerth was exploring underwater caves, she was doing research.

Louis Braille spent years and years researching, looking for an easier way for blind people to read.

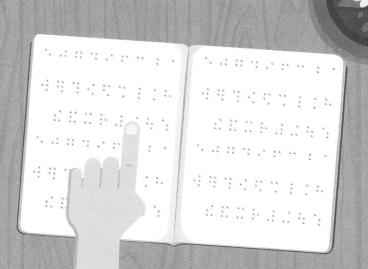

When Thomas Edison was trying to create the light bulb, he did a *lot* of research.

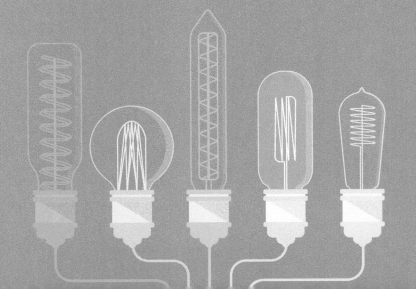

When Richie Stachowski wanted to find a way to talk underwater, he did some research. What did he do exactly?

First he looked on the internet and learned that sound travels better under water than out of water. Then he observed this for himself.

He put together something he thought would work under water and tried it out. This gave him information he could use to improve it, make it work better. This was useful research.

He kept doing this kind of research until he had something that worked well.

Some people think research is just about reading what other people have written. This can be part of research, but only a part!

In fact, research often means trying out ideas that are completely new. So it often starts with your imagination!

A hundred years ago, people thought the idea that humans could fly to the moon in a rocket ship was crazy! It was an idea imagined by a writer who loved thinking about what could exist in the future. He wrote a story about it. People read his story, and some began to think that someday it just *might* be possible for people to travel in space. They began to imagine too!

Eventually scientists began to research, experiment and learn about real space flight. And so today it *is* possible. And when you think about it, the idea first came from someone's imagination.

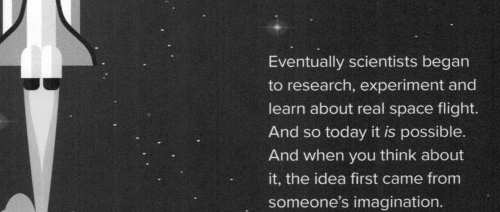

When you do scientific research, you get to do a lot of imagining. You get to look at things differently and think about things in new ways. You get to wonder what might be possible. You might well come up with an answer no one has ever thought of before.

This makes research challenging and fun!

Leonardo Da Vinci, who lived from 1452 to 1519, was a painter, scientist and inventor with a huge imagination. He dreamed of flying and drew plans for many flying machines, like planes and helicopters. He also designed parachutes, an armored car, a robot, scissors and the first diving suit.

When scientists first started researching how men could travel in space, they had to use the science they knew, but they also had to imagine what *might* work. Thousands of questions came up, and imagination played a big part in coming up with possible answers. Then experimenting showed which ones would work.

When you decide to research something, you will usually start by learning from others, perhaps by reading books or talking to other scientists. You can learn a lot that way.

But keep in mind that every great scientist has a great imagination.

without imagination and creative thinking, research is simply studying things that are already known.

You probably know a little about Benjamin Franklin. He is famous for many things, including a number of scientific inventions. In fact, he was a leading scientist of his time because of his experiments resulting in the discovery of electricity!

Franklin was an excellent observer and a sharp student. But he also had a wonderful imagination.

He imagined, then invented, a pair of glasses that would help a person see better both near and far. We now call these bifocals, where the top part of the lens helps you see far and the bottom part helps you see up close.

Ben Franklin also imagined, then invented, a type of fireplace called a "Franklin stove." At a time when people had only fireplaces to heat their homes, and rooms were often cold and full of smoke, the Franklin stove produced more heat, less smoke and was safer to use than a regular fireplace. It was a great improvement at the time, and it sprang from his great imagination.

When he was 11 years old, Ben loved to swim, but he wanted to swim faster. After some observation and research, he imagined a way he might be able to do this. To try his idea, he made two paddles out of wood and strapped them to his hands. It worked! He could swim faster! Then he tried a pair on his feet. They worked too. He had invented swim fins!

Sometimes research ends up giving us inventions that are just plain fun. Lonnie Johnson was a scientist who spent much of his life working with other scientists on spaceship design and travel. He won many awards for his valuable and important work.

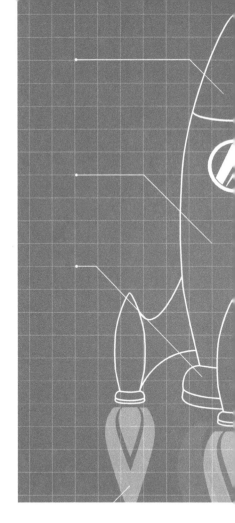

But he is best known for a more playful invention. He was working on a pump for a scientific invention when something unexpected happened and Lonnie's imagination quickly went to work. He described it like this: "I accidentally shot a stream of water across a bathroom where I was doing the experiment and thought to myself, 'This would make a great gun.'" And right there the Super Soaker, was invented!

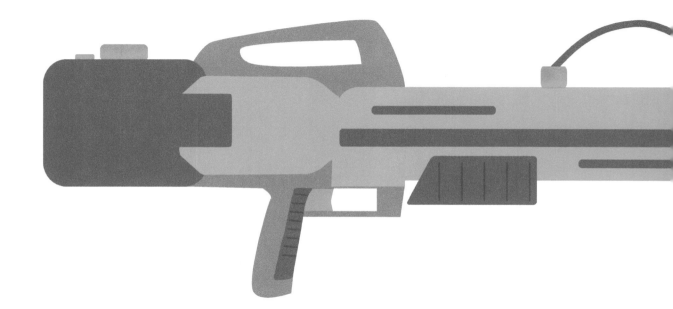

As you get better at doing research, just keep one thing in mind: Don't leave your imagination behind. It's one thing about you that has no limits!

RESEARCH

For this activity you will need

- your science journal

- a pencil

Steps

1 In your journal make a list of 10 things you'd be interested in researching.

2 Choose one that you can find out more about by observing directly, as well as by researching in other ways.

3 Use your imagination to come up with some questions, or things you'd like to know about it.

4 Have some fun using each of these kinds of research:

- reading something about it

- asking someone questions about it

- observing it closely

5 If possible, imagine an experiment you could try. Do it, if possible.

6 Put some notes of things you learned in your journal.

7 Share what you found out from your research with someone. Include any answers you found to the questions you asked.

HYPOTHESIS

When research and imagination come together, you end up with what's called a hypothesis (hy POTH uh sis).

A **hypothesis** is an idea you get about why something is a certain way or how something works. It's like a best guess that could lead you closer to an answer. A hypothesis normally comes after you've done some research and observation that has gotten you thinking:

"Hmmm. What I'm noticing here could be caused by...."

or

"Maybe this happens because...."

or

"The solution to this problem might be...."

George Washington Carver was a scientist who was interested in plants. He noticed that growing cotton year after year in the same field seemed to weaken the soil—the plants produced less and less cotton each year. This made life hard on farmers who made their living growing cotton.

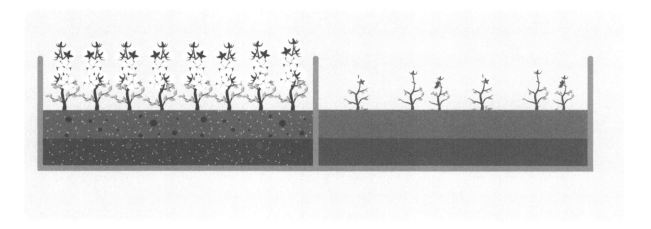

Carver was sure there must be a solution. He did a lot of research and used his imagination. He decided that growing different kinds of plants in the cotton fields might help strengthen the soil again. His hypothesis was that growing cotton plants year after year in the same soil had used up important nutrients. Not having these nutrients, the cotton plants were weak and didn't produce much cotton.

He thought perhaps a solution might be to use the field for different plants that would return those nutrients to the soil. If this worked, the soil would once again be full of the nutrients cotton plants needed, and they could once again grow healthy and strong.

He tried growing soil-enriching plants like peanuts, soybeans and sweet potatoes in the fields that had grown cotton. They grew well. After a few years of these crops, Carver replanted the fields with cotton. His hypothesis proved correct! The cotton grew well and the harvest was excellent.

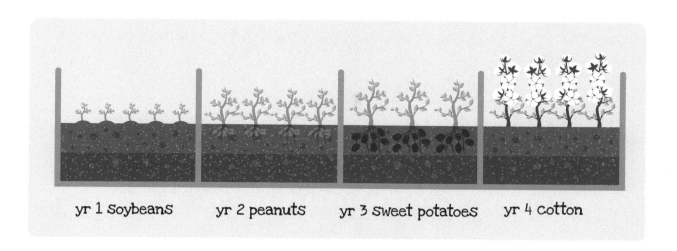

yr 1 soybeans   yr 2 peanuts   yr 3 sweet potatoes   yr 4 cotton

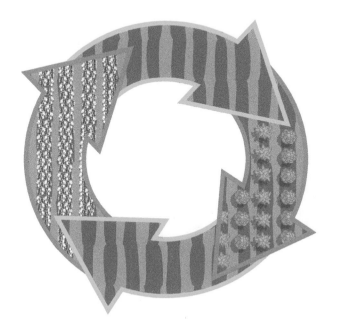

This method of keeping soil healthy is called crop rotation. Carver convinced many cotton farmers to rotate their crops this way. He even helped them with farming techniques and how to make money from the other crops.

In the end, his work helped many poor farmers improve the success of their farms. He helped them live better lives.

After you make a hypothesis, you need to test it to see if it is true. You need to find out if it works, if it leads to some improvement. This is what Carver did. He made a hypothesis, then he tested it.

Sometimes your testing shows you need to research and imagine some more. Sometimes scientists try one hypothesis after another, learning something each time. They don't worry about being wrong. They learn from it. This is all part of research.

Be curious, do your research, get imaginative and come up with a hypothesis. Who knows, you may come up with something that works!

The trick is to always keep your imagination at work.

Think like this:

Right or wrong, a hypothesis will help move you forward in your quest for discovery!

EXPERIMENT, EXPERIMENT, EXPERIMENT

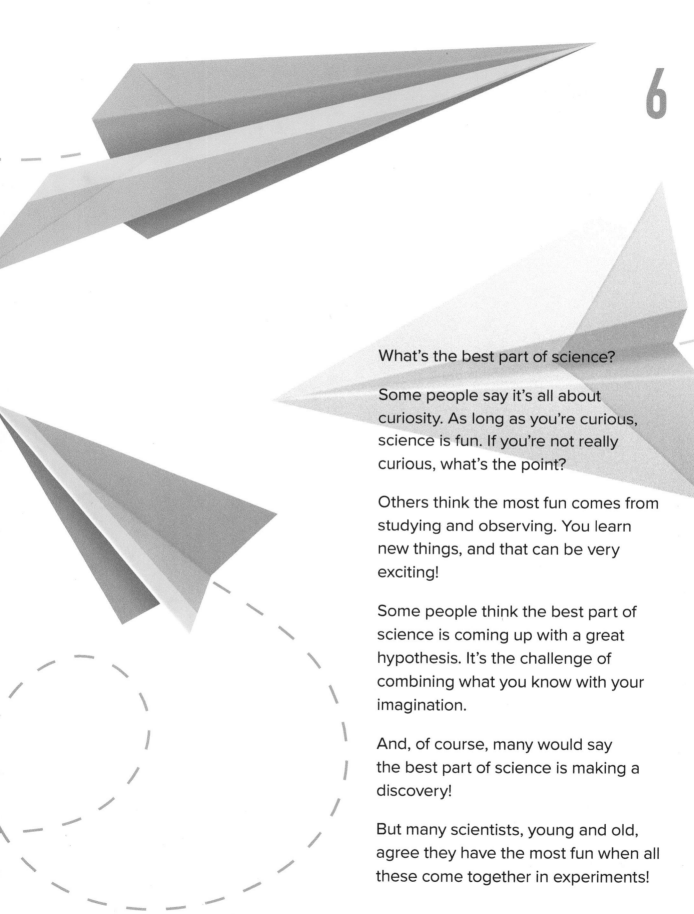

6

What's the best part of science?

Some people say it's all about curiosity. As long as you're curious, science is fun. If you're not really curious, what's the point?

Others think the most fun comes from studying and observing. You learn new things, and that can be very exciting!

Some people think the best part of science is coming up with a great hypothesis. It's the challenge of combining what you know with your imagination.

And, of course, many would say the best part of science is making a discovery!

But many scientists, young and old, agree they have the most fun when all these come together in experiments!

Why? Because an experiment is when you take everything you've learned and test your hypothesis to see if it works!

Doing research and learning about things can be exciting all by itself. But in the end, what good is all that knowledge if you can't do something with it?

That's where experiments come in. This is where you find out if you're on the right track.

Percy Julian was a scientist who researched and helped create a number of important medicines for treating problems with the human body.

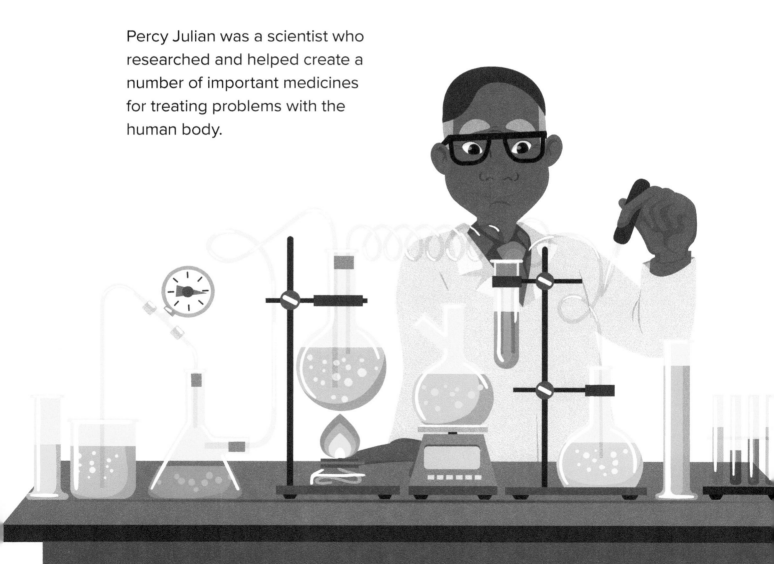

About a hundred years ago, he became interested in finding a treatment for an eye condition that caused blindness. Doctors had found a way to treat this, but the medicine they used came from the calabar bean and this only grew in a certain part of Africa. There just wasn't enough calabar bean to make all the medicine needed. So Percy decided to use his expert knowledge of chemistry to figure out how to create the medicine in a laboratory.

After three years and many different experiments, he was successful. Now plenty of medicine could be made and doctors all around the world could use it to help people keep their eyesight.

Percy Julian knew that a hypothesis always needs to be tested. He also knew that if you couldn't prove something with an experiment, you weren't done researching.

RESULTS AND CONCLUSIONS

Once you've done some experiments to test a hypothesis, you will have the results to look over. You expected certain things to happen. Did they?

Did something else happen that you didn't expect?

Was your hypothesis right, partly right, or was it wrong?

Looking over the results of your experiments, you make conclusions. **Conclusions** are decisions about whether your hypothesis proved to be right or not based on the results of your experiments.

Your conclusion might be that you need to research more.

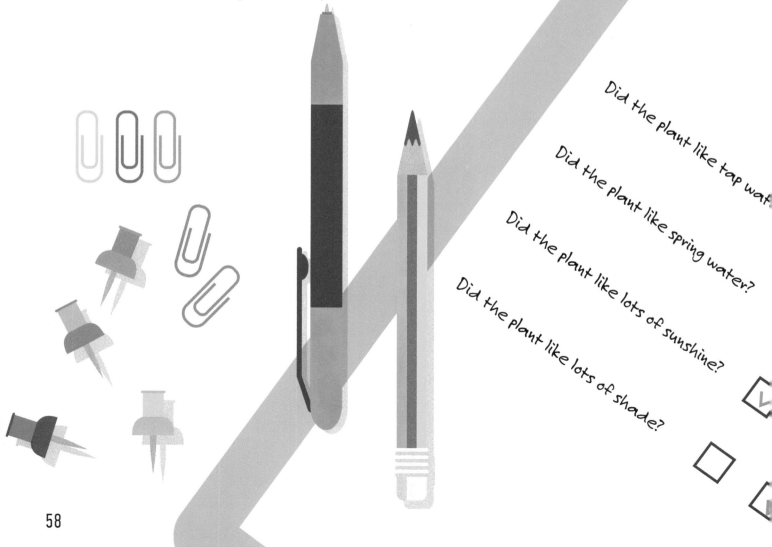

Did the plant like tap wat

Did the plant like spring water?

Did the plant like lots of sunshine?

Did the plant like lots of shade?

Or, depending on what you learned from your first experiment, you might decide to change your hypothesis and try a different experiment.

You might even decide to do more experiments to make very sure your hypothesis is right.

Whatever happens during an experiment, a smart scientist always takes good notes of the results and any conclusions made.

Yes

No

This way, whenever you want to look back at what you tested or what the results were, you have it written down so it's not forgotten. Keep in mind that you can also use videos, photographs and voice recordings to capture experiment results. These can be very helpful if you end up wanting to share your results with others.

ALL SHAPES AND SIZES

Experiments aren't always done in a laboratory or a classroom. They can be done anywhere. And they don't have to be fancy or complicated. Experiments come in all shapes and sizes, and sometimes they are very simple.

Here's an example of a simple experiment you can try anywhere.

Let's start with a simple hypothesis: *people like it when you listen carefully to what they are saying*. Maybe you already know this is true, but you can still do an experiment to test the hypothesis.

Talk to a few different people. Whenever they start talking, don't listen to them well and talk about something else instead. See what happens.

Then talk to a few more people and this time listen carefully to what they have to say. See what happens.

That's an experiment. It will prove that the hypothesis is either correct or incorrect.

Here's another example of a simple experiment.

Hypothesis: *caffeine makes a person's heart beat faster.* Again, maybe you already know this is true. But a simple experiment can prove whether it is or not.

Measure the heart rate (number of beats per minute) of someone else, or yourself. Make sure they haven't been doing anything active before taking their heart rate.

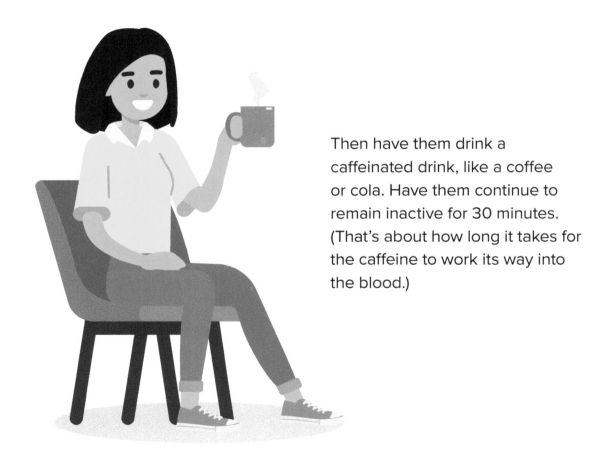

Then have them drink a caffeinated drink, like a coffee or cola. Have them continue to remain inactive for 30 minutes. (That's about how long it takes for the caffeine to work its way into the blood.)

Then measure their heart rate again. If you want, you can test it on more people or change some other part of the experiment to see what happens.

Whenever you are researching and come up with a hypothesis, the next step is to think up one or more experiments. That way you can test the hypothesis to see if it works.

Just remember: experiments come in all shapes and sizes.

And one of the great things about experiments is that they get you *doing things*. They allow you to test your ideas in the real world! You get to see whether they work, or not.

Did you have a hypothesis about something but never got to prove it? Then it might be time to design a good experiment and see what happens!

EXPERIMENT

For this activity you will need

- your science journal

- a pencil

- materials for an experiment

Steps

1 Choose a hypothesis to test. It could be one you thought of earlier, or a new one you make up for this activity. Write it in your journal.

2 Come up with an experiment that might help you prove or disprove your hypothesis. Write it in your journal.

3 Write down a plan for your experiment. Show it to an adult before starting, to make sure it's workable and safe.

4 Carry out your experiment. Make notes in your journal about what you did, what you observed, and what you learned from your experiment.

5 Was your hypothesis proven right? Congratulations!

Was your hypothesis proven wrong? Congratulations!

Either way, you learned something new. Now you have more information to use in coming up with the next experiment!

6 Is there a different experiment you'd like to do to further test your hypothesis? If so, note it down in your journal. Show it to an adult before starting, then do it.

OR

Do you want to change your hypothesis and test again? If so, note this in your journal, come up with an experiment, show it to an adult and do it.

Don't forget to keep good notes in your journal!

SHARING
KNOWLEDGE

7

When all your research and experiments are done, when the results are clear and you've made your conclusions, what's next?

Well, it depends on why you were researching in the first place. If you had just wanted to answer a question, you may be done.

Often, however, scientists want to share their results. This way others can learn from their work.

All the scientists we've talked about in this book were able to share the results of their work with others.

- Ben Franklin showed people how to make bifocals.

- Richie Stachowski got to share his Water Talkie and Lonnie Johnson got to share his Super Soaker.

- George Washington Carver shared what he learned about rotating crops with farmers, helping them be more successful.

- Louis Braille helped blind people all over the world, while Percy Julian shared his knowledge to help prevent people from losing their eyesight!

Perhaps you've heard of Lewis and Clark?

Meriwether Lewis was an American explorer and scientist who, with his friend George Rogers Clark, helped lead an exploration of America's west.

In 1803, President Thomas Jefferson asked Lewis and Clark to make a special voyage of discovery to explore and map the land, find out what kinds of plants and animals were there, and find a way to the Pacific Ocean. Off they went, knowing very little about where they were going or what they would find there.

Lewis was the group's chief scientist. He kept journals describing where the group went, what they found, what the weather was like, every new plant and animal they discovered, the people they met, what the land was like, and the rivers, lakes and mountains they found.

His journals contain lots of drawings of the new plants and animals they found, along with many maps of the new places they explored. He also made a collection of plant samples.

People still marvel at how many new things Meriwether Lewis observed and kept track of in his journals, and how much was learned when he shared everything he saw.

Many scientists write books or magazine articles so that others interested in their research can learn from them.

Often scientists get together at large meetings called conferences to share their knowledge. They talk about their experiments and results. They encourage other scientists to come up with new experiments to test their ideas and take them farther.

SCIENCE NEWS
SCIENCE WEEKLY
SCIENCE REPORT

As a young scientist, you may have the opportunity to share your knowledge with others as well. Many schools hold science fairs where students can describe their research, experiments and results. It's a great place to learn about all kinds of things—for students *and* adults.

Even if it's just telling a friend, it's always more exciting to be able to share what you've learned.

The world is full of knowledge. Share in it whenever you can!

Like all young scientists that came before, you play a role in keeping science moving forward. And as you become a better scientist, you will want to have practiced sharing what you learn. Someday your ability to share knowledge just might change the world!

SHARING KNOWLEDGE

For this activity you will need

- your science journal

- another person

Steps

1. Take a look through the notes you wrote when you did the experiment to test a hypothesis.

2. Share with someone what you learned from your hypothesis and experiment.

3. Find something else interesting that you put in your science journal.

4. Share it with someone.

THE SCIENTIFIC METHOD

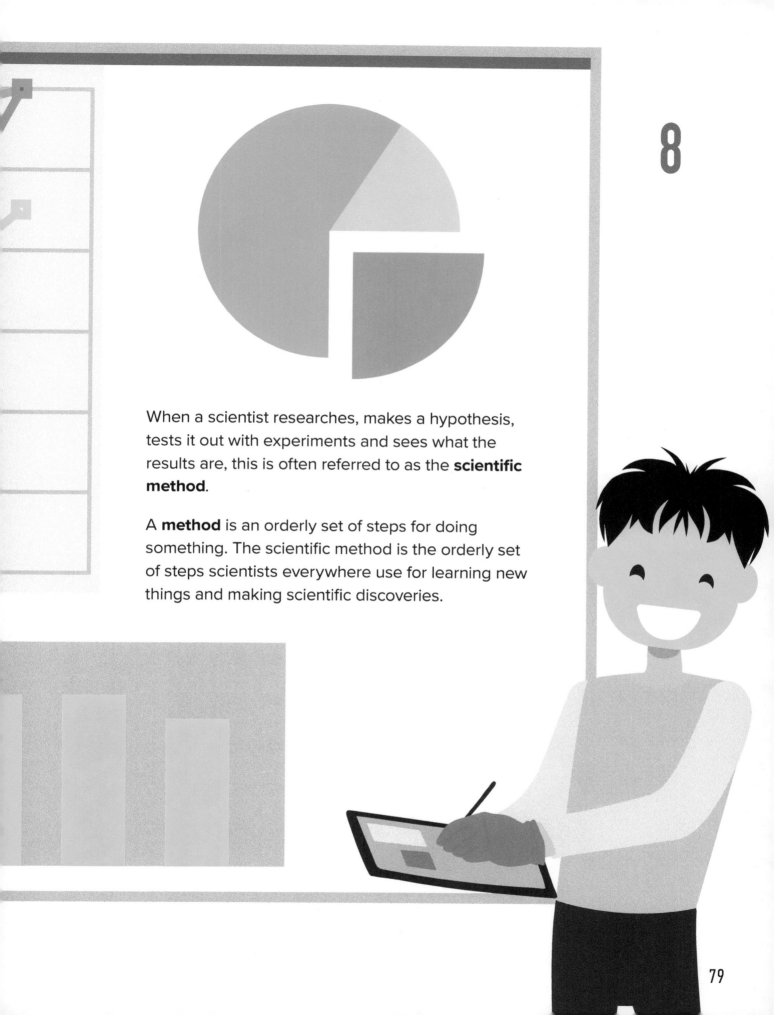

When a scientist researches, makes a hypothesis, tests it out with experiments and sees what the results are, this is often referred to as the **scientific method**.

A **method** is an orderly set of steps for doing something. The scientific method is the orderly set of steps scientists everywhere use for learning new things and making scientific discoveries.

The steps of the scientific method are usually written something like this:

1. Question

2. Research

3. Hypothesis

4. Experiments

5. Conclusions

6. Sharing results

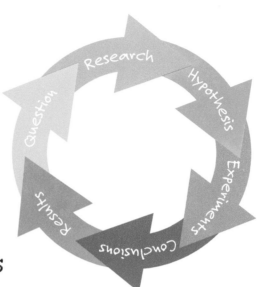

You'll see that these are all the steps we've talked about in this book. You might even look these steps over and think of naming them a little differently, for example:

1. Curiosity

2. Observation

3. Imagination

4. Testing

5. Results

6. Knowledge

As you gain experience doing research and making discoveries, you can decide for yourself how you want to describe the steps of the scientific method.

How will you know if you're doing the steps right? Ask yourself these questions:

- Am I getting results?

- Am I understanding the world better?

- Am I discovering things?

If the answers to these questions are "yes," then your scientific method is correct, because those are the only things that matter!

If you asked every one of the scientists we've learned about in this book, "What are the steps of the scientific method?" you would probably get different answers from every one of them.

The names of the steps aren't what is important. What's important is finding answers, solving problems, and maybe helping to make things better through science!

The right method is the one that works.

LEARN THE SCIENTIFIC METHOD

For this activity you will need

- your science journal
- a pencil
- another person

Steps

1 Learn these steps of the scientific method by heart. When you know them well, say them to another person and explain what each one means.

   a) Question

   b) Research

   c) Hypothesis

   d) Experiments

   e) Conclusions

   f) Sharing results

2 Learn these steps of the scientific method by heart. When you know them well, say them to another person and explain what each one means.

    a) Curiosity

    b) Observation

    c) Imagination

    d) Testing

    e) Results

    f) Knowledge

3 If you want to, make up your own version of the scientific method steps, one that you like. It can have fewer steps, more, or the same number. Memorize these. When you know them well, say them to another person and explain what each one means.

4 If you haven't already, write the version you like best in your journal.

WHERE
CAN
SCIENCE
TAKE US?

9

So, as you can see, science is all about understanding things. It's about finding out new things about the world and how it works.

And this means it's also about solving problems. Many of the problems we face every day can be solved by science. For example:

How could all the trash we throw away every day be turned into something useful?

How might the world grow more food so no one is hungry?

How could we cut down on pollution by inventing something to replace plastic?

How could we make batteries for electric cars that will run for hundreds of miles?

How can we predict earthquakes better?

At this very moment there are scientists working on every one of these problems, and many more. They are all working to make our world a better, happier place to live in.

At the same time, there is still so much to learn about the world and how it works, and so many important problems to be solved. How can we get every person on Earth a supply of safe drinking water? How can certain diseases be cured? Could humans live on any of the other planets in our solar system?

There are many different areas, or fields, of science to explore. Here are a few of the main ones.

BIOLOGY

Biology is the study of living things, both plants and animals. Within biology are many more fields. Some biologists study birds, others concentrate on insects, mammals or fish. Some study trees, some research microscopic life like cells, bacteria and germs.

MARINE BIOLOGY

Marine biology is all about the animals and plants that live in saltwater. Our planet's oceans are huge and scientists are still discovering new things about undersea life. Today, for example, a number of marine biologists are helping us learn more about coral and how to protect our coral reefs from pollution.

ZOOLOGY

Zoology is a part of biology that concentrates on animals of all kinds. A well-known zoologist you may have heard of is Jane Goodall. She spent many years living among the chimpanzees of Africa, studying their habits and helping the world understand all the ways chimpanzees use their intelligence and work together.

BOTANY

Botany is all about plants, and a scientist who studies plants is called a botanist. Remember George Washington Carver? He was a botanist. If you're interested in plants, how best to grow them or new and different ways they can be used, this might be a field for you.

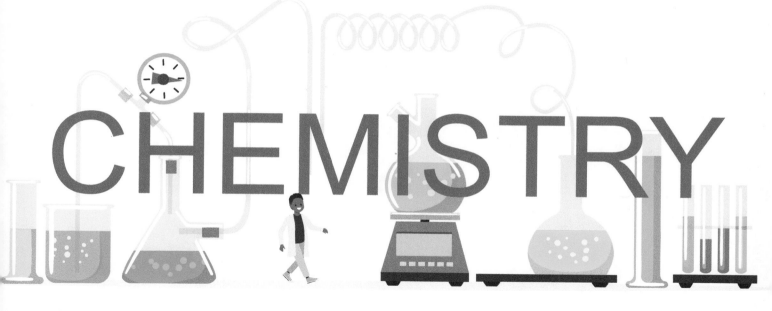

CHEMISTRY

There are scientists who know all about chemicals and how they work. **Chemistry** is the study of the chemicals everything is made from, for example oxygen, hydrogen and sodium chloride (the chemical name for salt). It's about what happens when you combine chemicals in different ways, and how to use them to make new substances.

In the 1800s the chemist Alfred Nobel invented dynamite. Among other things, this became very useful in building our roads, bridges, tunnels and railroads.

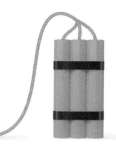

H Hydrogen + **O** Oxygen + **N** Nitrogen + **C** Carbon

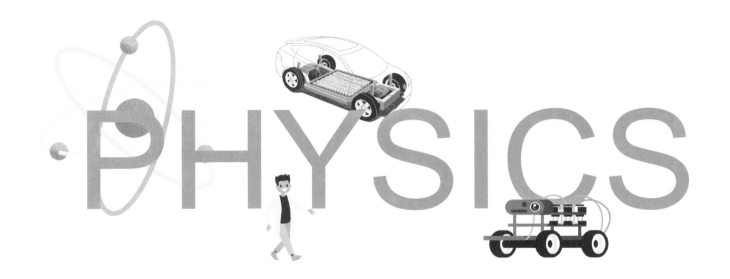

PHYSICS

In **physics**, scientists study matter and energy, light, heat, sound, and electricity. What is gravity? How do atoms work? How does light energy from the sun turn into heat energy on your skin? How do different kinds of motors work? These are all examples of things a physicist would study.

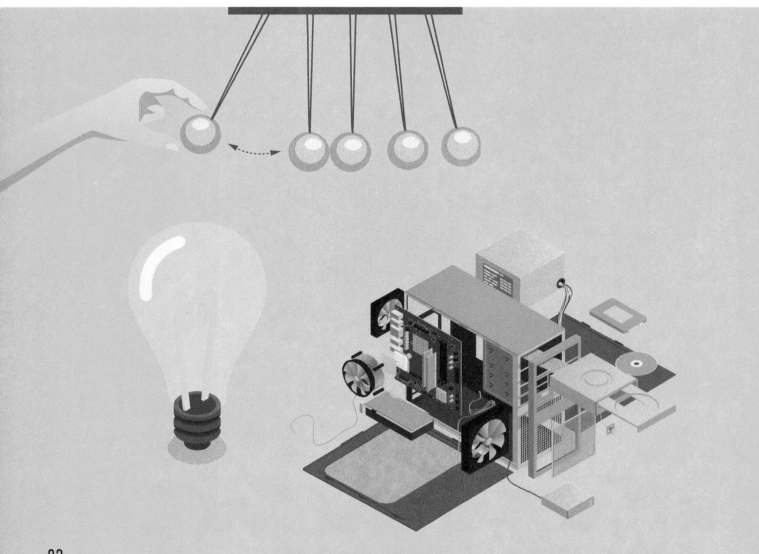

METEOROLOGY

You might think **meteorology** is about meteors, but it's not. A meteorologist is a scientist who knows all about the atmosphere, the layer of air surrounding our planet. Among other things, meteorologists use their knowledge of the atmosphere to predict our weather. Will there be a hurricane, or a tornado? A flood? Will it rain or snow soon?

GEOLOGY

Geology is the science that studies the earth, and the materials it's made of—rocks, metals, minerals and oil, for example. Geologists keep track of how the earth changes over time due to things like earthquakes, landslides, volcanoes, storms and floods.

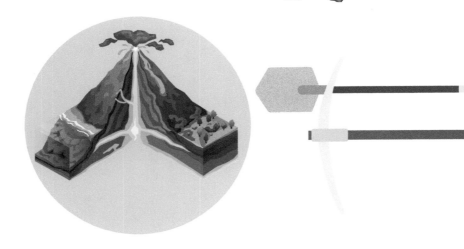

Geology also includes the study of other planets near Earth and what they are made of. The Mars rovers, for example, have been sent to gather information about the geology of Mars—what its surface is like, its rocks, craters, soil, volcanoes, water and minerals.

ASTRONOMY

And if you're interested in distant stars, suns, planets, galaxies, comets and other things in outer space, that's the field of astronomy. This field of science includes everything outside the earth's atmosphere, things we can see, as well as things we need huge telescopes and other equipment for. Astronomy has sparked the curiosity of humans since the beginning of time and there is still much to discover about the vast universe we live in.

With so many fields of science to explore—and these are just a few—you can see how science can take us anywhere our curiosity leads.

Some say that there is no problem in the universe that can't be solved by science.

What do you think?

FIELDS OF SCIENCE

For this activity you will need

- your science journal
- a pencil
- blank cards
- another person
- a small group of other people

Steps

1 Make a list of the fields of science discussed in Chapter 9. Think of, or find out about, 5-10 more fields of science and add them to the list.

2 Make a set of drill cards for your whole list of scientific fields. The front of each card should have the name of the field, and the back should have a short description of that field.

3 Practice these cards with another person until you know them well. When shown the card, you say the name of the field and explain what it is.

If they want, help the other person practice as well.

4 Pick one or two of these sciences you would like to find out more about, and do so.

5 Prepare a short presentation on what you found out. Present it to one or more other people.

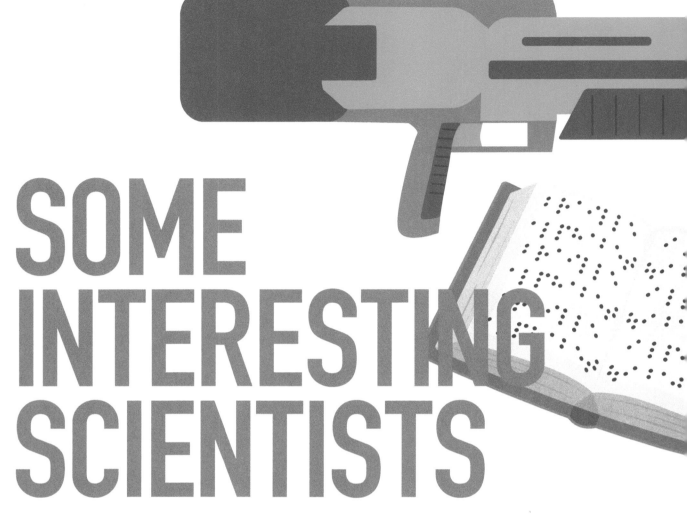

SOME INTERESTING SCIENTISTS

Through this book you've met some interesting scientists, from Percy Julian and Louis Braille to Jill Heinerth and Lonnie Johnson.

Here are a few more you might find interesting and inspiring.

Galileo, who lived from 1564 to 1642, was an Italian astronomer. He built several strong telescopes and used these to study the sky. He was the first to notice the moon had mountains and craters. Then he discovered the planet Saturn and its rings. He also discovered that the Milky Way was not just a cloud, as it appeared from the earth, but was a vast number of separate stars. And so he changed people's idea of the size of the universe!

About 300 years later, along came Edwin Hubble, an American astronomer. Back in the 1920s, only 100 years ago, people thought that our galaxy, the Milky Way, was the only galaxy that existed. Isn't that amazing? Hubble was able to use a huge telescope, one much stronger than Galileo's. He could see, and prove, that there were more galaxies out beyond the Milky Way. Once again a scientist changed our understanding of the universe!

Jacques Cousteau, a French diver, scientist, inventor and researcher explored what lay beneath the surface of the world's oceans. He invented the aqua-lung, an air pack with a breathing device, so divers could stay underwater for long periods. With this he was able to explore the exciting, beautiful underwater world that most people didn't even know existed. Cousteau researched, wrote about and filmed this world, and through many books and movies introduced millions of people to the beauty of life in our oceans.

Katherine Johnson was a mathematician who used her advanced math skills to calculate the flight paths for the first American space flights in the 1960s. John Glenn, the first human to orbit the earth in a spaceship, would only go into space if he knew that Ms. Johnson had checked all the math.

In 2015, at age 97, she was awarded the Presidential Medal of Freedom.

Steve Wozniak, a computer scientist, and his inventor friend Steve Jobs, are famous for co-inventing the Apple computer. Maybe you have used an Apple phone or tablet. Imagine starting the work to create these wonderful tools that are used by millions and millions of people.

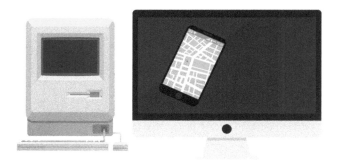

Timothy Berners-Lee is an English computer scientist who, in 1989, invented the World Wide Web! Then in 1990, he invented the first web browser. After that he created the very first website, which explained the World Wide Web and how to use it. His invention of the World Wide Web has been named one of the most important moments of the 20th century. It changed people's lives by making it possible to communicate instantly. What an amazing way to improve the world!

Suzanne Simard is a biologist who studies trees and forests. After years of research she had come to believe that the trees in forests form underground networks of fungus and roots which let them help one another grow. After a lot of experimenting, she found out this was true! Through this network of fungus and roots, trees can tell what nutrients other trees need and can then help them by sharing these nutrients.

Today, she thinks it's possible that trees can feel the presence of people. She is still doing experiments to find out if this is true. One day in the future we will find out if she's right about this interesting hypothesis. If it's true, imagine how we might be able to use this information!

All these successful scientists were curious. They had questions, things they didn't understand but wanted to know all about.

So they observed the world around them. They researched, sometimes quickly, and sometimes for years and years.

They used their imaginations. They tested their ideas. They often failed many times, but persisted. Eventually, they found answers to their questions. Eventually, they found solutions to the problems they had observed.

Their work has helped us understand more about the planet we live on.

It has improved the world for all of us!

Are you curious what
you might be able to do
in one or more fields
of science?

You're a young scientist.
Go find out!

Let's Do This!

IMAGINE A SCIENTIST

For this activity you will need

- paper (or science journal) and pencil or pen

Steps

1 Make up or imagine a scientist. Decide whether this scientist is from the past, the present or the future. Give them a name and an age. Decide what field of science the person worked in, what they were interested in, and what discovery or discoveries they made.

2 Write a brief description or a short story about your scientist, what he or she was like and what they accomplished.

3 Illustrate your story if you want to.

4 Share it with another person.

Printed in the USA
CPSIA information can be obtained
at www.ICGtesting.com
LVHW070729110224
771274LV00024B/131

9 780897 392464